Variations of the Scientific Age

Miguel A. Sanchez-Rey

Table of Contents

The Third Task

2-6

The Religious Naturalism of the Scientific Age

8-11

Destabilization of the Religious State and Ultranationalism

13-14

Logical Intuitionism

16-18

The Grand Unification Scheme and Metaspace

20-22

Decision-Making in the Scientific Age

24-26

Desperation and Totalitarianism

28-32

The Third Task

[Author: Miguel A. Sanchez-Rey]

The planetary super-state must be dismantled at the dawn of the Advance

Age to avoid a revolt against the state or from being seen as an external threat to an

extraterrestrial intelligence. The super-state exists in the Scientific Age as a

civilization that is on the verge of Class 2 and in that manner close to harnessing

the energy of a star only approximately 100 years after human civilization arrives

at the Advance Age. A peaceful civilization that has made First Contact with

countless alien civilizations in many different regions having settled in almost a

dozen habitable zones that will perpetuate human survival.

Unable to cohabitate very little will be known about these alien beings or

what their affairs may entail, but only that so few has ventured onto the outer

habitable zones to explore and associate with these diverse alien races in a galaxy

teeming with primitive life in which only so few has made it as far as Class 2.

Even then humanity dares not encroach into their solar habitats and other habitable

areas they lay claim. Choosing only to promote and exert the internationalist

model that will safe-guard human self-interest and security from any impending

hardship.

Humanity is gripped by the economics of mutual self-interest and markets.

They are associationists' in free-trade and the self-management of labor. Under the jurisdiction of the council of the scientific planetary super-state. Whom must oversee the authoritarian control and the defense of the planetary super-state for the welfare and tranquility of the workforce and its democratically elected workers' councils. That is the council of the planetary super-state unilaterally upholds and protects the democratic order from any existential external/internal threat.

The beginning of the Advance Age anticipates a transitional period in which the planetary super-state is dismantled. That is anticipatory existence becomes a necessary prerequisite for a healthy transition where decision-making in matters of economics is made in ever increasing small steps. Taking short-strides, at a growing rate, in decision-making allows the self-management of labor to make exact long-term outcomes without finiteness.

For unfeasibility is known to be the problematic conundrum of perfect anticipation. In which anticipatory economic existence is the near-perfect anticipation of both flaw and precise decision-making. Where if the super-state is to be dismantled then contact will be lost between Earth and the outer habitable zones.

If that's the case, then perfect decision-making becomes harder to achieve and in that flaw decision-making becomes, yet again, a dire threat. For if Class 2 anticipates the achievement of the harnessing of a star, and Anarcho-syndicalism must successfully take-over the role of the declining religious state, then human beings must interconnect decision-making in such a way that flaw-decision making is near avoidable and peaceful co-existence is achieved that is long-lasting and wildly strong in its capacity to prolong humanity's perpetuation until the Omega-Kardeshev scale reaches finitude.

The Third Task poses the final question for PHPR [The Physicalist Program]: how does economic existence avoid futility in ever increasing steps until flaw decision-making is overtaking by anticipatory existence? In which avoiding futility is a paradoxical apparition to anticipatory economic existence.

It's a mathematical science task that anticipates incalculable formalism and ingenuity. Ushering in the dismantling of the planetary super-state and the strong-independence of the space-habitats, terraformed planets, and the regional industrial federation of planet Earth and the outer habitable zones.

Where the internationalist model is actualized and the habitable zones give way to Anarcho-syndicalism. A highly advance technological civilization which, by harnessing the energy of a star, accomplishes the unforeseeable. By also harnessing Incalculability, they achieve perpetual anticipatory existence.

The Religious Naturalism of the Scientific Age

[Author: Miguel A. Sanchez-Rey]

Religious naturalism believes that the natural world can serve as both a religious and scientific paradigm of rational existentialism and metaphysical Platonism. That the loss of conception of authority in the historical consciousness, as an unnecessary component of Western thought, gives way to a wildly strong predisposition for logical intuition. That is religious naturalism holds truths to be both foundational and empirical. In which the interplay seeks to guide religious naturalism as an umbrella theology of religious world-views and praxis. That is celebratory in its tradition of social openness and its acceptance of racial differences and sexuality.

The experimentation of the modern era is to self-actualize liberal theology and in which liberal theology seeks to make self-empowerment as an embodiment of the declining religious state even as authoritarian practices are to be enacted to mitigate social and psychological pathology and degeneracy. But in which authoritarian practices are to neither inhibit nor suppress human potentiality and self-utilitarianism.

A naturalism which in its norm believes that Anarcho-syndicalism is the logo-centric outcome of the religious experience. For which the religious experience stipulates the unanimity and unification of emotional and psycho-cognitive attachment to natural law theory. And in which in its contract holds democratic self-ownership to be a necessary component of personal legality and felicity. Though naturalism is in itself antithetical to axiomatic truths, rational existentialism helps to bind axiomatic truths with naturalism by abandoning being as a lost cause in the phenomenological reduction. For which reductionism is only one of many strategies in the interplay of ideas.

That is Anarcho-syndicalism is the superior outcome of the religious naturalist movement where the conception of authority is lost and in which the religious states decline does not lead to suppression of self-actualization and social cohesion. Self-interests is an intricate role of evolutionary biology and in which evolutionary epigenetics plays a primary role in the rising force of Anarcho-thinking.

That is Anarcho-syndicalism is seen as the federation of industrial economies. By which religious naturalism is understood as an umbrella theology. For which religious-syndicalism can be acknowledge as the umbrella theology of the federation of industrial economy. That is the dismantling of the authoritarianism of the super planetary-state, so it does not lead to the totalitarianism of the super-state, will lead to the self-management of labor and the anticipatory economics of the internationalist model.

Where the semantic connotation of religious-syndicalism embodies the disintegration of normative beliefs of authority toward one of self-empowerment and self-actualization. That is in which self-reliance is a social endeavor but in which all social groups strive to reciprocate to meet their individuality.

Though the religious state is seen as a sacred institution in which human rights are protected and enforced, there is, nevertheless, the decline of the religious experience of the state. In which political disenchantment and collective psychosis has rendered the religious state ineffective as the human population is self-aware of the collapse of social democracy and desires only for the scientific socialism of the state but in which the state is antagonistic to their preconceptions of natural rights. Power-structures are to be dismantled when their use is no longer of any value. And in which values are a necessary component of human well-being and in which human life is a sacred part of existentiality. That is well-being is secured by self-management and human life is motivated by self-actualization.

Destabilization of the Religious State and Ultra-nationalism

[Author: Miguel A. Sanchez-Rey]

The religious states decline, due to a disenchantment with the political process and a world-wide meltdown cause by the collapse of social democracy, has cause a flight-or-fight response to preserve what's left of the religious state within certain groups and political interests. Leading to the rise of ultra-nationalist sentiment within those circles. But in which ultra-nationalism implies resistance against the inevitable outcome and the internationalist model. Where it's futility is tantamount as disenchantment has cause the religious state to destabilize due to its sudden decline.

Destabilization which has resulted in the inadequacy of state services and functions design to protect the population from the ravishes of private enterprise. Leading to desperation where the only alternative is to incite nationalist fervor without encouraging radicalism and extremism. But in which radicalist and extremist are inciting opposition to the state, under the cover of distributive justice, in order to build nationalist opposition to the establishment in the formality of popular politics.

Logical Intuitionism

[Author: Miguel A. Sanchez-Rey]

Logical intuition posits that one logically intuits to arrive at a logical conclusion. That is one gathers information intuitively and logically sorts out that intuitive information, using logical inference, deduction, induction, and etc., to the extent that a logical conclusion can be made about the information that is asserted. Information that if applied in any other way may yield flaw argumentation that can mislead or hinder progress in natural epistemology. That is if:

$$P \rightarrow Q$$

$$\therefore Q \rightarrow R$$

$$P \rightarrow R$$

Which implies that if P has truth value, Q has truth value, then R has truth value. But arriving at R, which has truth value, then there exist an S such that:

$$\exists Q \exists P : P \wedge Q \rightarrow R$$

$$\therefore \exists S \exists R : R \rightarrow S$$

17

$$P \wedge Q \rightarrow S$$

$$\Rightarrow Q \rightarrow S$$

$$\Rightarrow P \rightarrow S$$

Which is understood as a primary example of logical intuitionism. That is one sorts out logically, in a step-by-step fashion, what one intuits to be true and/or what one intuits to be false until a conclusion is made S that can be gathered by breaking statements into its atomic form such that a theorem can be stated and a proof is devised that holds its logical conclusion valid.

If one were to intuit logically to arrive at an intuitionist logical conclusion, then one will may arrive at an intuitionist conclusion that has no bearing such that a flaw premise or statement may jeopardize, disqualify or cast into doubt its intuitive conclusion.

The Grand Unification Scheme and Metaspace

Miguel A. Sanchez-Rey

The Physicalist Program

Abstract

Advance superstrings are consider incalculable particles. With this in mind a more refine definition of the grand unification scheme and metaspace is presented.

July 15th, 2017

The Definition of the Grand Unification Scheme:

The collection of advance superstrings that exist in $\varnothing \subseteq H_M$ that can related and interchanged through their super-charge monopoles on the 11-supermembrane manifold.

> **Metaspace**:
>
> **Cosmological homotopic states between advance superstrings.**

Decision-Making in the Scientific Age

[Author: Miguel A. Sanchez-Rey]

With the chaos brought on by the LIGOS experiment, that drove the scientific process over the edge with the search for dark matter, one has reach the Scientific Age in the form of an uncontrollable scientific machine. In that manner, with the Brexit in the European Union and the 2016 American Presidential Elections, the planet went into a global meltdown consistent with a schizophrenic breakdown that led to collective psychosis.

The social justice movement, striving to protect the democratic ownership of the state and the neo-Fascist movement that strive to uphold the primacy of corporate capitalism, soon morph into a neo-Fascist alliance that overtook the political process.

With the collapse of social democracy, in which the religious state began to permanently decline after the 2016 Presidential Elections, as one of the worst crimes in world history, decision-making, both in political and economic policy, came to a halt at a world-wide scale that has ravish the House of Congress into a debating body and has resulted in deadlock in the general elections in the United Kingdom.

Inevitably leading to a world-wide epidemic of rising ultra-nationalist sentiment with all other alternatives, as well, unable to sustain the social contract that upheld the democratic ownership of the state. In which the mental-thought processes of planetary society become more graphic and intense.

Causing the magnification of the state that will ultimately lead to the founding of the planetary super-state. To preserve what's left of the religious state, it must resort to authoritarian practices through democratic decision-making. That is the religious states decline led to a global shock in which decision-making, both in politics and the international markets, become more centralized and in which political and economic decision-making, on common grounds, becomes, as well, a hindrance to state-capitalism and neo-Keynesian economics.

A result of the inevitable fact that popular policy is undesirable on both ends of the political spectrum. Where if popular policy were to be the norm a scientific dictatorship would arise but in the end the futility of the state, subjected to destabilization, will lead to more centralized political and economic decision-making but in which the self-management of industry pacifies the general population from revolting against what's left of the religious state. Desperation subsides and decision-making becomes a volatile reality in the Scientific Age.

Desperation and Totalitarianism

[Author: Miguel A. Sanchez-Rey]

Totalitarianism is a desperate act to control the population when neither conflict nor plenty can withstand the revolt of the masses. The totalitarian state then resorts to terror politics to solidify its control of the general population by utilizing brutal tactics of control that keeps them in line with the ruling class. It's then that the totalitarian state becomes an existential problematic to peaceful co-existence, for which the dictatorship of the proletariat posits the transition into anarchy when unnecessary power-structures are dismantled in a gradual process that starts with a revolt against the bourgeoisie, assimilates private enterprise into state ownership of industry, and finally the gradual dissolution of power-structures that leads to the realization of anarchism politics. But also in which the dictatorship of the proletariat disregards self-interests and self-actualization.

That is neither the state nor private interests will interfere in the self-actualization of the historical consciousness into a materialism that is the fruition of communism. Communism which stipulates, through classical liberalism, that man is to pursue work for its own creative ends and that centralized power-structures can inhibit the full potential of self-development and self-actualization. That is the state is the antagonist and man is the protagonist.

The totalitarian state suppresses individual self-actualization and self-empowerment. Where total control of state services and functions is enacted in that which political society is subservient to the totalitarian regime or ruler. Whereby the totalitarian state implements desperate policies to keep the population in line with the central power-structure and in that in which external and/or internal conflict can no longer motivate the population.

Therefore, desperation is inherently the nature of the rise of totalitarian ideology. An ideology in which total control is a consequence of a pathological mass movement. A pathological mass movement that enforces terror and violence to control the population in totality and so that the population does not revolt against the totalitarian regime. A last resort to realize a pathological political ideology that has no bearing on human rights or tranquility. That is antagonistic to natural rights and is favorable to one or more groups over others.

It is in this sense that desperation breeds into the totalitarian ideology as the mass-movement of control-freaks and thugs whom can no longer preserve their positions in power but must then resort to terror policies of brainwashing, eavesdropping, massacring, and all other war-crimes, to keep the population in line with the state. A state that suppresses individual self-actualization and self-empowerment. Where total control of state services and functions is enacted in that which political society is subservient to the totalitarian regime or ruler. Whereby the totalitarian state implements desperate policies to keep the population in line with the central power-structure and in that in which external and/or internal conflict can no longer motivate the population.

A predisposition for conflict emerges and that in which the conflict between the state and the individual leads to a subtle realization that both incentives and sacrifices are to be made to keep the individual in line with the state. But since state worship is an intolerable aspect to individuality, and that in which the state is a sacred institution that protects man from the state-of-nature, then the individual sees the state as both a protagonist and antagonist.

Therefore, the resolution, to both the protagonism and antagonism of the state, is the super-state. A state in which the self-management of industry elects its own workers' councils and its own workers' council selects those amongst each other to take part in the scientific council of the super-state. Whereby the scientific council of the super-state takes unilateral action to protect the democratic order and resorts to authoritarian practices to keep the general population in line with what's left of the religious state by democratic decision-making.

In which the dismantling of the religious state will lead to dissolving the council of the super-state and eventually all that remains is the self-empowerment and self-management of the federation of industries. Where man is now free to take part in the scientific process, free from bondage to the state, and wildly strong. That is the end result of perpetual anticipatory economics is a healthy transition into an Anarcho-syndicalist stage.

Where Anarcho-syndicalism is the response to an incalculable desperate act. In that there's no other alternative. And that which any other alternative could yield immeasurable harm to the democratic principle of self-empowerment and the human right of self-actualization. For which wild-strength stipulates that man measures itself rather than to be an immeasurable objectification. So that the incalculable desperate act is a necessary precondition to a resolution to the antagonism of the state and the protagonist of the individual. For which the individual asserts itself as the dominion regime of the state, and that in which the nation-state is the dominion of the collective consciousness, Anarcho-syndicalism thrives in the collective interface. So that the collective interface postulates the non-interference and non-mutilation of individual awareness and consciousness. And that totalitarianism is a horrid act of a belittling nature.